GLOBAL WARMING, OUR REAL ENEMY: What You Need to Know About China's Worst Heat Wave.

LIANG YÜAN

Table of Contents

Introduction
Chapter 1
Chapter 2
Chapter 3

Introduction

The world's worst heatwave has been recorded in China. Energy, water supplies, and food production are all suffering in China as a result of an extended period of extremely hot and dry weather.

People, industry, and agriculture are all being affected by the record-breaking heat and little rainfall that has affected most of China. Large tracts of crops have been damaged, enterprises have closed due to electrical shortages, and river and reservoir levels have dropped. The issue might have global ramifications, further upsetting supply systems and escalating the international food crisis.

Two months of intense heat have been felt by residents of huge portions of China. Numerous records have been broken and temperatures of over 40°C (104°F) have been recorded in hundreds of locations.

People can cool off in the rest spaces that have been set up in subway stations.

The highest temperature ever recorded in China outside of the arid region of Xinjiang was 45°C (113°F) on August 18 in Chongqing, Sichuan province. The greatest minimum temperature ever recorded in China in August was reached on August 20 when the temperature in the city didn't drop below 34.9°C (94.8°F). A maximum of 43.7°C (110.7°F) was recorded.

The highest temperature ever recorded in China outside of the arid region of Xinjiang was 45°C (113°F) on August 18 in Chongqing, Sichuan province. The greatest minimum temperature ever recorded in China in August was reached on August 20 when the temperature in the city didn't drop below 34.9°C (94.8°F). A maximum of 43.7°C (110.7°F) was recorded.

Since national records began in 1961, this heatwave in China has been the longest and hottest on record. It is the worst heatwave ever recorded, according to weather historian Maximiliano Herrera, who tracks extreme temperatures around the globe.

He claims that this simultaneously combines the highest intensity, the longest length, and the largest possible area. "Nothing in the history of the climatology of the world is even similar to what is happening in China."

In addition to the high heat, portions of China's limited rainfall have caused rivers to drop to low levels, with 66 of them drying up. The Yangtze has some of its lowest water levels since records have been kept, which was in 1865. A few locations have run out of their local water sources, necessitating the delivery of drinking water by truck. For the first time in nine years, China issued a national drought emergency on August 19.

The decreased hydroelectricity production is a result of the low water levels. Because Sichuan typically receives 80% of its electricity from hydropower, it has been particularly impacted. Due to a lack of electricity and the enormous demand for air conditioning, thousands of firms in the province had to close their doors. To conserve energy, offices and retail centers were also instructed to turn down the lighting and air conditioning.

There have reportedly been agricultural losses of 47,000 hectares and crop damage of 433,000 hectares in Sichuan alone. According to the agriculture ministry, cloud seeding will be used to attempt and boost rainfall. The impact of cloud seeding is still debatable from a scientific standpoint.

China is not the only country experiencing drought. The current drought in Europe could be the worst in 500 years. Along with parts of the US and Mexico, the Horn of Africa is experiencing a drought.

Reduced crop production in these areas could make the world food problem worse. Even before Russia invaded Ukraine, food prices reached record highs; even though they have since declined, they are still higher than in previous years. However, China has recently accumulated significant grain reserves, so it can make up for some of the loss.

Droughts have been increasing as a result of global warming, and as the earth continues to warm, they will become more common and severe, according to a 2021 report from the Intergovernmental Panel on Climate Change.

Chapter 1

What is Going on in China?

The official duration of the current heat wave in China has been 68 days, making it the longest heat wave on record since the National Climate Center of China began keeping data in 1961. This week, forecasts for temperatures above 104 degrees Fahrenheit were made by more than 240 localities. That is quite sexy. The first national drought alarm in nine years has been issued by the nation. Additionally, it has pushed some regions—like central Hubei Province, for instance—to quite severe measures. They are so desperate for rain that they announced this week that they are beginning to seed clouds, an experimental technique in which metal is sent into the clouds. You, therefore, believe that it may rain. They need water because the scarcity is also harming other areas' power supplies, but this isn't verified.

Despite the lack of rain, a region in the southwest named Sichuan relies on hydroelectric power to the tune of around 80% of its energy needs. And now is the season when there is a significantly larger demand for energy because everyone is using air conditioning to stay cool. With over 80 million residents, Sichuan chose last week to turn off the electricity to most factories for around seven days to guarantee that everyone would have access to electricity at home and prevent the power grids from becoming overloaded.

The bad news is that Sichuan is home to several sizable global corporations' plants. Foxconn is a producer of electronic components. There are American businesses with manufacturing that are very dependent on reliable power, like Intel and Texas Instruments. Additionally, these manufacturers work together to produce vital electronics and automotive components. Therefore, even though the electricity will only be down for a week, it will take much longer for them to resume production when it does. And that will further exacerbate the current global scarcity of semiconductors.

Furthermore, China already faces economic challenges as a result of these intermittent COVID restrictions. And the current heat wave is not helpful. And the most recent economic statistics for the nation show that this impact has already been felt. They were released last week. They're really bad. They reveal that consumer spending fell short of all projections. Officially, there are around 20% of youth without jobs. Home sales have decreased, and the real number is likely far higher.

They claim that they will make an effort to strike a balance between the COVID regulations, the current heat wave, and the pursuit of economic growth. However, they have already lowered their yearly goals.

The goal of China's central bank's interest rate cut is to make small firms more affordable to lend to, so lessening the impact of the COVID and heat wave. However, analysts claim that won't help much. Things appear scarier in the long run. This summer's events in China are probably going to become more frequent. A large area of northern China may become uninhabitable by the end of this century due to excessive heat, according to a 2018 analysis from the Massachusetts Institute of Technology. And because the earth cannot absorb water when it does rain, flash flooding, on the other hand, is becoming a more frequent occurrence. Furthermore, there were two flash floods in China this week alone that claimed 22 lives.

One nation outperforms them all in a year where climate records for floods, fires, and drought are being broken every day like shards of glass: China. An unprecedented heatwave is sweeping the nation.

With record-breaking daytime and nighttime temperatures, it has continued for more than seventy days.

In a region of the country covering 500,000 square miles, more than 100 million people are impacted by record high temperatures of over 104°F (40°C). This is comparable in size to the combined areas of Texas, Colorado, and California.

The "worst heatwave ever recorded in global history" or "worst heat wave known in world climatic history" is how experts are referring to it.

They are not the only ones who have described the magnitude of what is occurring in China at this time. The severe event that is baking southern China "has no counterpart in modern record-keeping in China, or anywhere else around the world for that matter," according to Axios.

"I can't think of anything equivalent to China's heat wave of summer 2022 in its blend of intensity, duration, geographic range, and several impacted individuals," meteorologist Bob Henson, a contributor to Yale Climate Connections, told Axios.

In certain areas of the nation, temperatures have reached 110.3°F (43.5°C), breaking existing records. The 9 million-person city of Chongqing set another record last weekend with a nighttime low temperature of 94.8°F (34.9°F).

Rivers that were once vast and powerful, like the Yangtze, the third-largest river in the world and the source of drinking water for 400 million people, are now mere shadows of what they once were. With levels, more than 50% below the average for the past five years, several sections of the river are closed to shipping. Some rivers have been dried up to mere trickles and puddles.

According to analysts like Jefferies, China's record-breaking two-month heatwave was a "black swan event" since COVID-19 was still present and still having an impact on the nation. According to Jefferies, the drought has produced a vicious cycle in which the lack of hydroelectric power reduces the industry's ability to produce electricity.

Additionally, the country's reduced reliance on hydropower as a result of the drought has resulted in blackouts and electricity shortages. According to reports, Sichuan, which receives more than 80% of its energy from hydropower, is currently experiencing a "terrible scenario," for which the government has declared a serious "level 1" emergency.

China issued a formal statewide drought alert last Friday as the government increased measures to combat the heat. The lack of drinking water in many locations has forced the government to deliver water or generators to those without either.

The country's upcoming ten-day rice harvest could be impacted by the drought, which is anticipated to extend far into next month. Due to the low lake levels, the old irrigation channels are ineffective, thus the officials are busy digging new ones.

And right now, the Chinese government is in such a state of desperation that they are attempting to create rain by "cloud seeding," a method of weather modification that entails spraying chemicals on clouds in the hopes that they will produce rain, in areas of the central and southwest of the nation.

Stopping oil and gas drilling is one method of ending the drought. Because specialists do not doubt that this is a result of climate change; The longest continuous stretch of high temperatures and little rain, "with the broadest in reach," was recorded in Southern China.

It has been consistently hot and dry in Southern China for the longest time since 1961, "with the largest in breadth and the highest average intensity."

Authorities in China have issued an urgent call for action to protect crops in the face of the nation's warmest summer on record, warning that the autumn harvest is under "serious threat" from high temperatures and dryness.

This summer's record temperatures, flash floods, and droughts have taken a toll on the second-largest economy in the world, phenomena that experts have predicted will occur more frequently and intensely as a result of climate change.

According to the agriculture ministry, Southern China has had the longest stretch of consistently high temperatures and little rain since records have been kept, which dates back more than 60 years.

On Tuesday, a notification was released by four government agencies requesting that "every unit of water" be conserved to safeguard crops.

According to the statement, "the quick development of drought combined with high temperatures and heat damage has produced a grave threat to autumn crop production."

China produces more than 95% of the rice, wheat, and maize it eats; nevertheless, a smaller harvest could result in a rise in import demand in the world's most populous nation, placing additional strain on already-stressed global supplies due to the crisis in Ukraine.

Many Chinese provinces have implemented power cuts due to temperatures as high as 45 degrees Celsius (113 degrees Fahrenheit), as towns struggle to meet an increase in demand for electricity that is partially fueled by people turning up their air conditioners.

Outdoor ornamental lighting has been reduced in the megacities of Shanghai and Chongqing, while Sichuan province has implemented industrial power cutbacks as a result of low water levels at important hydroelectric plants.

According to China News Service, a state-run media outlet, the sweltering heat is also drying up the crucial Yangtze River, with water flow on its main trunk about 50% lower than the average over the last five years.

In certain areas of the nation, authorities have already started using cloud seeding, a technique to create rain.

This month, state television CCTV aired a video of firemen delivering water to thirsty farms and meteorology personnel launching catalyst rockets into the sky.

According to climate science, excessive heat is becoming worse by an exponential factor, she stated. Therefore, the likelihood that next year will shatter records for heat is higher.

Extreme weather has been referred to as nature's "revenge" on humans by government climate specialist Zhou Bing, who issued a warning over the weekend about mass migration brought on by climate change.

Three other periods of extreme heat have occurred in China so far this century: 2003, 2013, and 2017.

According to Zhou, the time between heatwaves is "substantially shrinking."

According to Xu, a native of Chongqing, "life continues with some fortitude" for those who have to endure the oppressive heat.

Chapter 2

Influences on Power

In China's Sichuan area, which depends on hydropower for 80% of its energy requirements, oppressive heat and dry conditions have made an already severe drought worse. As a consequence, the province, which has a population of over 80 million, was obliged to halt companies and request that citizens use less electricity.

Authorities prolonged the Sichuan power outages on Sunday until August 25. Meanwhile, other areas began their power outages. Due to their indiscriminate nature and the inability of factory managers to find workarounds, the effects on supply chains and the Chinese industry could be disastrous.

Mirko Woitzik, the worldwide director of intelligence solutions at Everstream

Analytics, a provider of supply-chain insights and risk analytics, told Fortune that "the situation has become worse over the last few days." The severity of the COVID-19 shutdown in Shanghai/Kunshan earlier this year was comparable to the scene last week, which was one of the worst weeks for industry closures in China.

According to Woitzik, restrictions on the amount of power that can be generated have caused delays at more than a dozen vehicle manufacturers, impacted the production of lithium, and hampered the processing of semiconductors at important facilities.

In Shanghai, the largest financial center of China, authorities said in notice over the weekend that buildings along the Yangtze River would not be illuminated on Monday or Tuesday to preserve energy, resulting in the absence of the city's iconic skyline.

According to Woitzik, there is another unseen bogeyman out there that is having an even greater influence on China's economic development and supply chains than these power outages to factories and office buildings.

"Transportation will also be hampered if temperatures do not decrease in the next few days," he added. "For example, through the Yangtze River, which is equivalent in its supply chain significance to the Rhine River in Europe, adding to industrial supply chain issues."

Supply lines in Europe were recently impacted by low water levels in the Rhine River, which hindered cargo boats from completing cargoes. Even Deutsche Bank referred to the river as the bloc's possible "Achilles' heel" as it battles an enduring energy crisis. China can see a similar predicament if the country's drought doesn't end soon.

Geographically, China's heat wave has expanded: official media reported that during the previous 70 days, temperatures exceeding 104° F (40° C) were recorded at least once throughout almost 530,000 square kilometers of the nation. The situation has become so terrible that over the previous two months, more than 260 Chinese meteorological stations have registered their highest-ever temperature readings.

According to meteorologist Bob Henson, "I can't think of anything equivalent to China's heat wave of summer 2022 in its combination of severity, length, geographic range, and several people afflicted."

Woitzik emphasized that the heat wave and dryness also led the Hunan area, which is located immediately to the east of Sichuan, to declare a state of emergency. The province relies on hydropower for around

40% of its energy demands, and as reservoir levels drop, the area is implementing power outages as Sichuan did to save energy.

Due to the drought and intense heat, China's biggest freshwater lake in the Hunan province has shrunk by 66%, exposing astonishing views for airline passengers traveling over the area.

That's bad news since Hunan produces 22% of China's rice, which is the primary diet for 65% of the people. The province is also well known for being a center for the production of specialized metals and equipment.

Chapter 3

How to be Secure and Comfortable During the Heatwave

Although most of us like warm weather, there are health dangers when it becomes too hot. More individuals than normal get extremely sick or pass away during heatwaves. Make sure the summer's heat doesn't affect you or anybody you know if it does.

Why are heat waves problematic?

A heat wave primarily puts people at risk for:

- not getting enough water (dehydration)
- being too hot, which may exacerbate symptoms in those who already have heart- or breathing-related issues

- heat exhaustion and heatstroke

Who is most in danger?

Anyone may be impacted by a heatwave, however, the following groups are especially at risk:

- People over 75, typically women, who live alone or in care facilities who have significant or chronic illnesses such as heart or lung diseases, diabetes, renal disease, Parkinson's disease, or certain mental health issues.

- Those who take numerous medications, who may be more adversely impacted by the heat.

- Persons who could have trouble staying cool, such as infants and the very young, bedridden people, those who are addicted to drugs or alcohol, and people who have Alzheimer's disease.

- People who spend a lot of time outdoors or in warm environments, such as those who live in high-floor apartments, are homeless or have outdoor occupations.

Advice on Surviving the Heat

If you can, avoid the heat. If you must go outdoors, remain in the shade, particularly between the hours of 11 a.m. and 3 p.m.; wear sunscreen, a hat, and light clothing; and keep away from exercise or other activities that increase your body temperature.

You should calm down. Eat and drink cold things, stay away from hot and alcoholic beverages, and take a chilly shower or splash cool water on your skin or clothing.

Make sure your home is cool. When it is dark and the temperature outside has dropped, open the windows that were closed throughout the day. If the temperature is below 35 degrees, electric fans may assist. Verify the room's temperature, particularly in areas where individuals who are more vulnerable reside and sleep.

If you believe a hot home is harming your health or the health of another person, you may also request assistance from the environmental health office at your local council. They may check a rental property for health risks, such as high heat.

Keep an eye out for symptoms of heat sickness. Heat exhaustion or heat stroke may be the cause of elevated body temperatures if you or another person experiences them in hot weather.

Learn the symptoms of heat exhaustion and heat stroke, as well as when to seek medical attention.

www.ingramcontent.com/pod-product-compliance
Lightning Source LLC
LaVergne TN
LVHW020540160826
845677LV00015B/4151
9798848280982